Ocean Animals

Whales

BLASTOFF! Beginners

by Christina Leaf

BELLWETHER MEDIA
MINNEAPOLIS, MN

Blastoff! Beginners are developed by literacy experts and educators to meet the needs of early readers. These engaging informational texts support young children as they begin reading about their world. Through simple language and high frequency words paired with crisp, colorful photos, Blastoff! Beginners launch young readers into the universe of independent reading.

Sight Words in This Book

a	have	it	out	to
are	help	jump	the	use
big	how	make	these	water
do	in	not	they	what
find	is	of	this	with

This edition first published in 2021 by Bellwether Media, Inc.

Library of Congress Cataloging-in-Publication Data

Names: Leaf, Christina, author.
Title: Whales / by Christina Leaf.
Description: Minneapolis, MN : Bellwether Media, Inc., 2021. | Series: Ocean animals | Includes bibliographical references and index. | Audience: Grades PreK-2
Identifiers: LCCN 2020007738 (print) | LCCN 2020007739 (ebook) | ISBN 9781644873298 (library binding) | ISBN 9781681038162 (paperback) | ISBN 9781681037929 (ebook)
Subjects: LCSH: Whales--Juvenile literature.
Classification: LCC QL737.C4 L323 2021 (print) | LCC QL737.C4 (ebook) | DDC 599.5--dc23
LC record available at https://lccn.loc.gov/2020007738
LC ebook record available at https://lccn.loc.gov/2020007739

Editor: Amy McDonald Designer: Andrea Schneider

Printed in the United States of America, North Mankato, MN.

Table of Contents

Whales!

What jumped
out of the water?
It is a whale!

Whales are big. Blue whales are the biggest animals alive!

humpback

beluga

Whales swim far.
They travel
to find food.

Body Parts

Whales have **blowholes**. They take in air. They blow air out.

blowholes

Whales have **flippers**. They help whales swim.

flipper

Some whales
have teeth.
They bite food.

teeth

Some whales
do not have teeth.
They trap food
with **baleen**.

baleen

Whale Sounds

Whales
make sounds.
These whales
use **echoes**
to hunt.

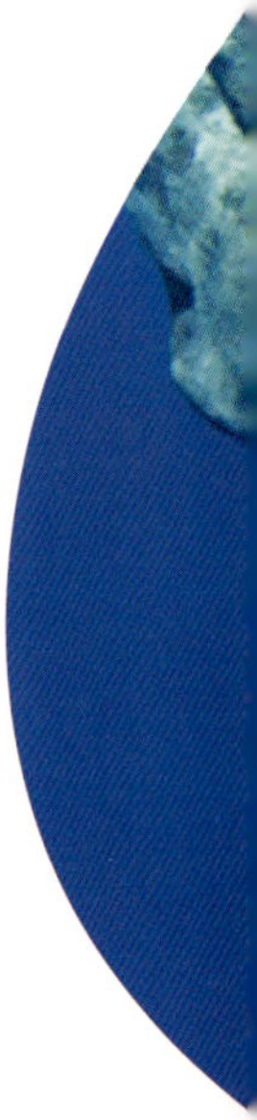

This whale sings.
This is how it talks.
Helloooo!

Whale Facts

Whale Body Parts

Whale Food

squids

krill

fish

Glossary

baleen

hard parts in the mouth that catch food

blowholes

holes that take in and blow out air

echoes

sounds that bounce off objects

flippers

wide, flat body parts whales use to swim

To Learn More

ON THE WEB

FACTSURFER

Factsurfer.com gives you a safe, fun way to find more information.

1. Go to www.factsurfer.com.
2. Enter "whales" into the search box and click 🔍.
3. Select your book cover to see a list of related content.

Index

The images in this book are reproduced through the courtesy of: Gudkov Andrey, front cover; WaterFrame/ Alamy Stock Photo, pp. 3, 6 (blue); Maria T Hoffman, pp. 4-5; Chase Dekker, pp. 6-7; Igor Kruglikov, p. 7 (humpback); Luna Vandoorne, p. 7 (beluga); Mathias Berlin, pp. 8-9; Cavan/ Alamy Stock Photo, pp. 10-11; redbrickstock.com/ Alamy Stock Photo, pp. 12-13; CJ Larsen, pp. 14-15; Pushish Images, pp. 16-17; wildestanimal, pp. 18-19; Nature Picture Library/ Alamy Stock Photo, pp. 20-21, 22 (baleen); Sean Steininger, p. 22 (parts); Lauren Squire, p. 22 (squids); Apple Pho, p. 22 (krill); Datenschutz-Stockfoto, p. 22 (fish); Forrest9, p. 23 (baleen); Ignacio Salaverria, p. 23 (blowholes); Miles Away Photography, p. 23 (echoes); Nico Faramaz, p. 23 (flippers).